MIMIC MASTERS!

By Tracy Vonder Brink

TABLE OF CONTENTS

A Crabtree Seedlings Book

Crabtree Publishing

crabtreebooks.com

School-to-Home Support for Caregivers and Teachers

This book helps children grow by letting them practice reading. Here are a few guiding questions to help readers build comprehension skills. Possible answers appear here in red.

Before Reading:

- What do I know about this topic?
    - *I know that some animals act like other animals to scare away predators.*
    - *I know that some birds are good at copying sounds they hear.*
- What do I want to learn about this topic?
    - *I want to learn which animals use mimicry.*
    - *I want to learn why mimicry is helpful.*

During Reading:

- What have I learned so far?
    - *I have learned that mimicry is an adaptation.*
    - *I have learned that adaptations help living things survive.*
- I wonder why...
    - *I wonder why a corpse flower often only blooms every two to three years.*
    - *I wonder why a female northern mockingbird chooses a mate with the most songs.*

After Reading:

- What did I learn about this topic?
    - *I have learned that a four-eyed butterflyfish has a spot on its tail to fool predators.*
    - *I have learned that a bee orchid's flowers look and smell like a female bee.*
- Read the book again and look for the glossary words.
    - *I see the word **survive** on page 3 and the word **mate** on page 4. The other glossary words are found on pages 22 and 23.*

MIMIC MASTERS!

An **adaptation** is something that helps a living thing **survive**.

Looking or acting like something else is called **mimicry**. Being a mimic is one adaptation.

These animals and plants are mimic masters!

A male northern mockingbird uses mimicry to find a **mate**.

He sings songs that sound like other birds.

A female chooses the male with the most songs.

A clearwing moth has stripes like a bee or a wasp.

Other animals think the moth can sting so they stay away.

A clearwing moth has no stinger. It only looks like a bee or a wasp.

A four-eyed butterflyfish has a spot near its tail.

The spot looks like an eye.

A **predator** cannot tell which way the fish is swimming.

A mimic octopus folds its arms against its head.

It copies the shape of a **poisonous** flatfish.

Predators leave the octopus alone.

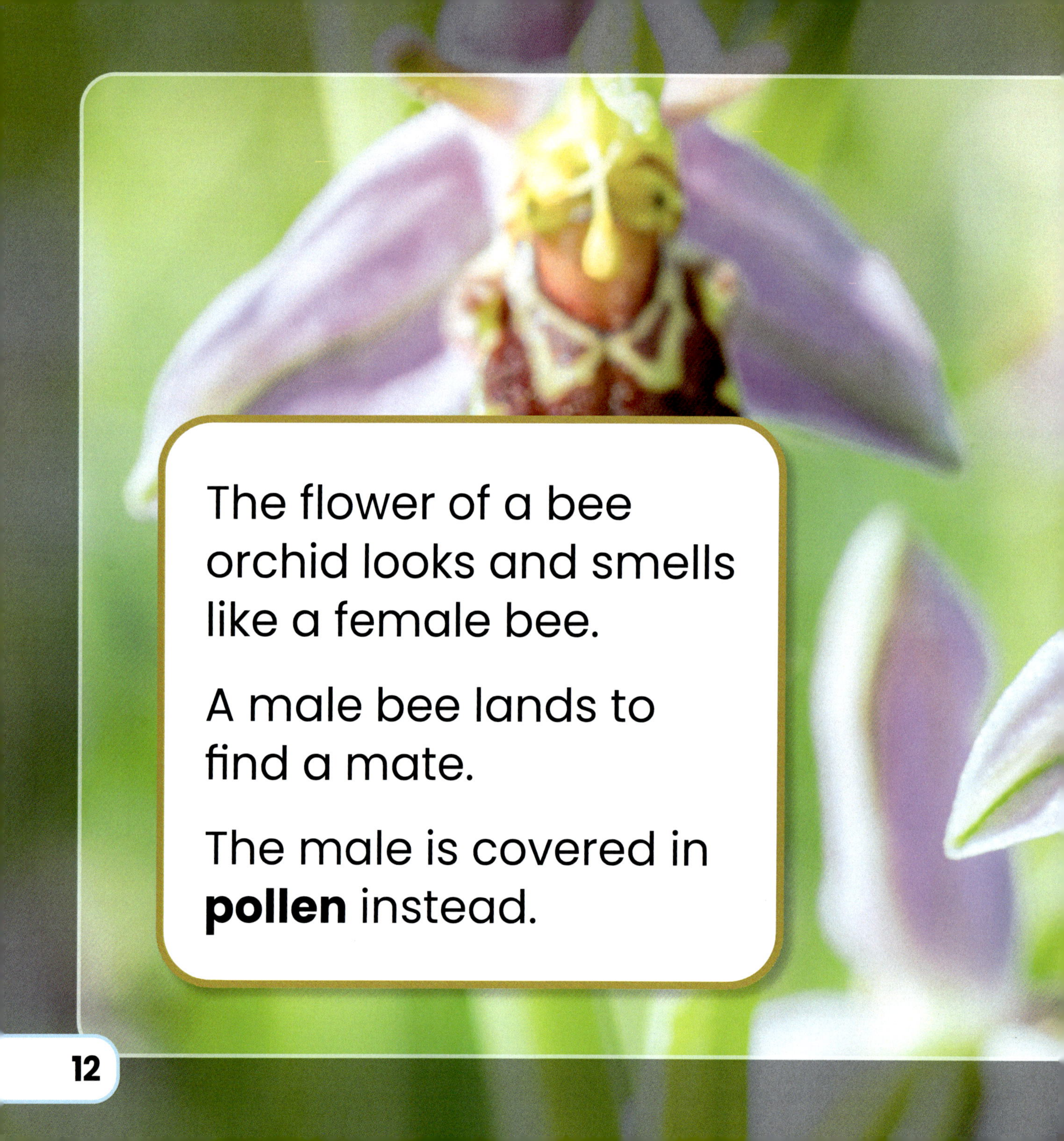

The flower of a bee orchid looks and smells like a female bee.

A male bee lands to find a mate.

The male is covered in **pollen** instead.

The male bee carries the pollen on its body to the next bee orchid. Plants use pollen to make seeds.

A corpse flower uses flies to carry its pollen.

It mimics the smell of rotten meat so flies will find it.

A corpse flower often only blooms for a few days once every two to three years.

An orchid mantis eats butterflies.

But first, the butterfly must come close enough to catch.

The mantis looks like a flower to attract the butterfly.

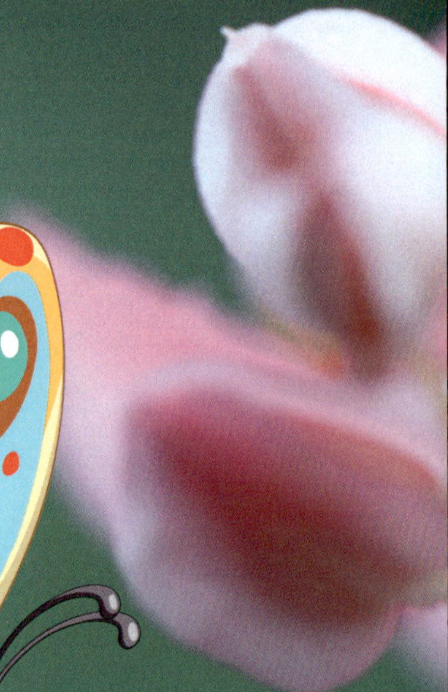

One part of a hairy frogfish's fins looks like a worm.

The frogfish wiggles it.

The frogfish catches shrimp that come to eat.

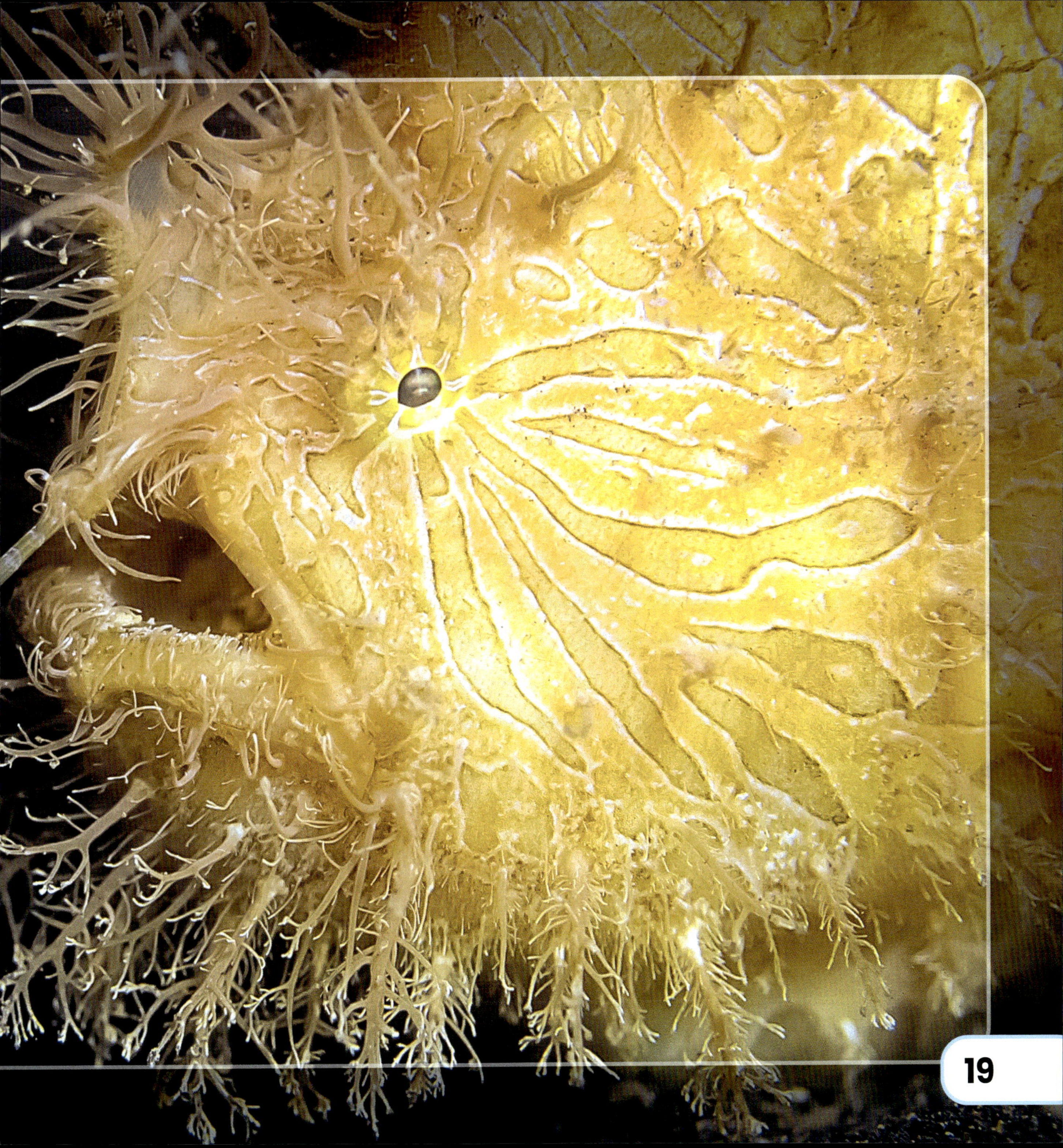

A female Eastern firefly flashes the pattern of a different firefly group to attract males.

But she doesn't want a mate. She wants to eat them.

Mimicry is a helpful adaptation for many living things!

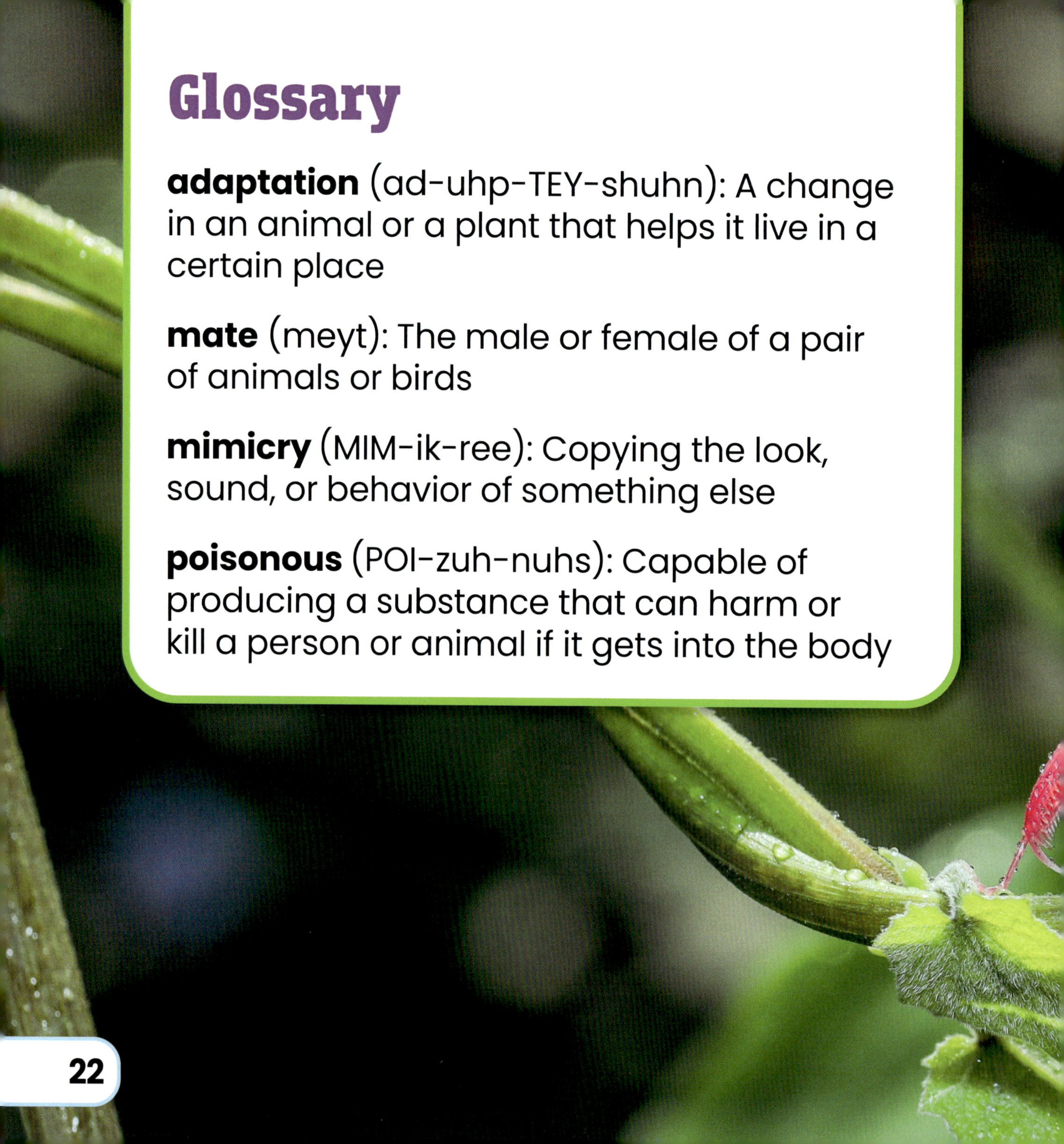

Glossary

adaptation (ad-uhp-TEY-shuhn): A change in an animal or a plant that helps it live in a certain place

mate (meyt): The male or female of a pair of animals or birds

mimicry (MIM-ik-ree): Copying the look, sound, or behavior of something else

poisonous (POI-zuh-nuhs): Capable of producing a substance that can harm or kill a person or animal if it gets into the body

pollen (POL-uhn): A powdery substance produced by flowers in order to create new flowers

predator (PRED-uh-ter): An animal that lives by hunting and eating other animals

survive (ser-VAHYV): To continue to live

Index

About the Author

Tracy Vonder Brink loves to learn about science and nature. She has written more than 100 nonfiction books for kids. Tracy lives in Cincinnati, Ohio, with her husband, two daughters, and two rescue dogs.

Written by: Tracy Vonder Brink
Designed by: Margaret Salter
Proofreader: Melissa Boyce
Photographs: All images from Shutterstock

Crabtree Publishing

crabtreebooks.com 800-387-7650

Printed in the U.S.A./CP2025

Published in Canada
Crabtree Publishing
616 Welland Avenue
St. Catharines, Ontario
L2M 5V6

Published in the United States
Crabtree Publishing
347 Fifth Avenue
Suite 1402-145
New York, New York, 10016

Library and Archives Canada Cataloguing in Publication
Available at Library and Archives Canada

Library of Congress Cataloging-in-Publication Data
Available at the Library of Congress

Hardcover: 978-1-0398-8111-2
Paperback: 978-1-0398-8472-4
Ebook (pdf): 978-1-0398-8231-7
Epub: 978-1-0398-8351-2
Read-Along: 978-1-0398-8556-1
Audio: 978-1-0398-8514-1